Curious Kids Guides
TIME AND SEASONS

Brenda Walpole

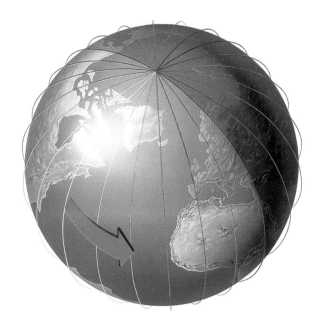

KINGFISHER

NEW YORK

KINGFISHER
a Houghton Mifflin Company imprint
215 Park Avenue South
New York, New York 10003
www.houghtonmifflinbooks.com

First published in 1996
First published in this format 2002
10 9 8 7 6 5 4 3 2 1
1TR/0502/TIMS/*UD UNV/128MA

LIBRARY OF CONGRESS CATALOGING-IN-PUBLICATION DATA
has been applied for.

ISBN 0-7534-5471-8

Printed in China

Series editor: Clare Oliver
Series designer: David West Children's Books
Author: Brenda Walpole
Consultant: Ian Graham
Editor: Claire Llewellyn
Art editor: Christina Fraser
Picture researcher: Amanda Francis
Illustrations: Susanna Addario 10-11; Peter
  Dennis (Linda Rogers) 4-5; Chris Forsey
  cover, 6-7, 20-21; Terry Gabbey (AFA Ltd)
  16-17, 24-25; Craig Greenwood (Wildlife Art Agency) 12-13;
  Christian Hook 26-27; Biz Hull (Artist Partners) 14-15, 22-23,
  30-31; Tony Kenyon (BL Kearley) all cartoons; Nicki Palin
  18-19; Ian Thompson 8-9, 28-29

CONTENTS

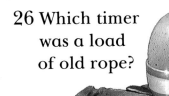

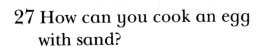

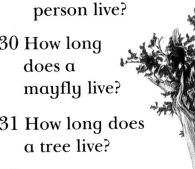

Why does the Sun rise in the morning?

The Sun doesn't really rise at all! It's the Earth which turns to give you a sunrise each morning. The Earth is like a spinning ball. Wherever you are, it starts to get light as your part of the Earth spins around to face the Sun. The sky grows brighter— and it's morning!

● Every morning, the Sun rises in the east. Its light wakes animals early. A new day has begun.

● The ancient Greeks believed that the Sun was a god called Helios. He rode across the sky in a chariot of flames.

● The weather may be gray and gloomy, but above the clouds the Sun is always shining.

Why does it get dark at night?

The Earth spins day and night. During the day, the Earth's movement makes the Sun seem to travel across the sky. The hours pass, and your part of the Earth turns farther away from the Sun. The Sun "sinks" in the sky and darkness comes. It's night.

● It takes 24 hours for the Earth to spin around once. While one half has daylight, the other half has night.

● At night, one side of the Earth is in deep shadow. This makes it feel much cooler, so you often need to wrap up warmly.

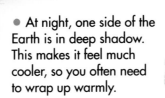

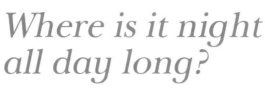

Where is it night all day long?

During the winter months, the lands around the poles don't see the Sun at all. The Sun is so low in the sky that it is hidden below the horizon. This makes the days cold and dark—even at noon.

• After the dark days of winter, it must be wonderful to see the Sun again. The Inuit people of North America used to celebrate its return by lighting new lamps in their homes.

• In the summer, everything changes at the poles. The Sun is in the sky morning, noon, and night. If you live there, it must be like sleeping with the light on.

● Winter days are dark in the northern lands of Scandinavia. Sami, or Lapp, children go to school by the light of the Moon and stars.

● Some people are so miserable in the dark days of winter that they become ill. Doctors find that bright lights often cheer their patients up.

Where do all the days start on time?

In tropical lands near the equator, the Sun rises at almost the same time every morning. And it sets at almost the same time every evening! The days and nights are each about 12 hours long—every single day of the year.

Why do we have seasons?

We have seasons because of the way the Earth moves around, or orbits, the Sun. Each orbit takes a year from start to finish. The Earth is tilted, and as the planet orbits, first one pole and then the other leans toward the Sun. This is what gives us the seasons.

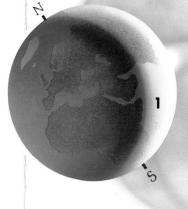

4 In September, neither pole leans toward the Sun. It is fall in the north, and spring in the south.

1 In December, the North Pole leans away from the Sun. The northern half of the Earth has winter. The southern half has summer.

● The Earth spins around as it orbits the Sun. It doesn't spin upright, though, but tilts to one side.

● The seasons are different on opposite sides of the Earth. Depending on where you are, wearing a bathing suit in December could give you frostbite or sunburn!

3 In June, the North Pole leans toward the Sun. The northern half of the Earth has summer. The southern half has winter.

● The weather is always hot at the equator because the Sun is high overhead. But at the poles the Sun is lower in the sky. Its rays are spread out and feel much weaker.

2 In March, neither pole leans toward the Sun. It is spring in the north, and fall in the south.

Why do we plant seeds in spring?

Seeds need to be warm and wet before they sprout. As the spring sunshine starts to warm the ground, farmers and gardeners dig the soil and sow their seeds. The plants won't take long to grow.

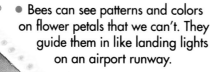

● Bees can see patterns and colors on flower petals that we can't. They guide them in like landing lights on an airport runway.

Why are bees busy in summer?

On warm sunny days, bees are busy visiting hundreds of flowers. Inside each flower is a drop of sugar-sweet nectar. Bees feed on this and use it to make honey back at the hive.

When do leaves drop off ?

In the fall trees find it difficult to
suck up water from the cold ground.
So their leaves dry out and turn
red, gold, and brown. They drop
to the ground, leaving the trees bare
for winter. The trees will grow
new leaves in the spring.

● Not all trees lose their
leaves. Conifers are trees
with tough leaves that
can stand the cold in
winter.

Why do animals sleep
through the winter?

For some animals, sleeping is the best way
to survive the hungry winter days.
Chipmunks, squirrels, and some
bears eat as much as they can
in the fall, then sleep
somewhere safe until
springtime.

● Many animals grow
thick winter coats to help
them survive the
bitter cold.

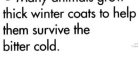

Where are there only two seasons?

Many tropical countries have only two seasons in the year. One is very wet and the other very dry. Not many trees manage to survive the dry months, and animals travel hundreds of miles searching for food and water.

● Many animals migrate in different seasons. Every year, swarms of monarch butterflies leave Mexico and fly 2,000 miles to spend the summer beside cool Canadian lakes.

● During the dry season, the ground is baked hard by the hot sun. Clouds of dust cover everything and everyone.

● In the dry season, herds of wildebeest and zebra cross the grasslands of central Africa. They follow the thunderclouds in search of rainwater and fresh grass.

EQUATOR

● Tropical lands lie near the equator. They are the warmest parts of the Earth.

Where does it pour for a month?

Some parts of India and Southeast Asia have long, heavy downpours called monsoons. Big black clouds are blown in from the sea during the summer months. Once the rain starts, it can last for weeks, flooding the fields and streets.

Who can tell the time without a clock?

We all can! Inside every one of us there is something we call our body clock. It wakes us up every morning and tells us it's breakfast time. And all through the day we seem to know just when it's time to work, eat, and play. As evening comes we feel tired and get ready to sleep.

• Newborn babies don't know their days from their nights. They just wake up their parents whenever they feel hungry!

• Different kinds of animals have different timetables. A bee and a badger never meet. One is active by day; the other is active by night.

Can flowers tell the time?

A few flowers are such good timekeepers that they open at the same time every day. Gardeners sometimes plant flowers like these in flower clocks. There are as many as 12 different flowers in the "clock" and they open one after another as the hours pass.

● Animals have body clocks, too. In zoos and on farms, many of them know when it's feeding time.

Can animals tell the time?

Some animals are active by day; others wake at night. Still others seem to know when the seasons change. The snowshoe hare's body reacts to the cold when winter approaches. It grows a white fur coat which makes it hard to see against the snow. This protects it from its enemies.

Which calendar was carved in stone?

Many hundreds of years ago, people called the Aztecs lived in Central America. They made a calendar from a huge stone shaped like the Sun. The face of the Sun god was carved in the middle, and signs for the days were carved all around the edge.

Who invented our calendar?

More than 2,000 years ago, a Roman ruler named Julius Caesar invented the calender we use today. He gave each year 365 days, and arranged them into 12 months. Since then, the calendar has hardly changed.

- The Aztec calendar stone measured nearly 13 feet from side to side—that's far too big to hang on the wall!

What is a leap year?

Every four years we have what we call a leap year. This is a year with 366 days instead of 365. The extra day is added to the end of February. So if your birthday falls on February 29th, it's a very special day indeed.

- A leap year can always be divided by four with none left over. The years 1996, 2000, and 2004 are all leap years.

Why do we need calendars?

Most of us need calendars to help us remember all the things we plan in a year. But they also help us keep track of time. When people are shipwrecked or taken hostage, they find unusual ways to mark the passing days.

Who lights the lamps at New Year?

The Hindu New Year is called the Festival of Light—and no wonder! Towns are strung with lights, and lamps shine from every door and window. The women make beautiful floor pictures with colored chalks, flour, and sand, and decorate them with glowing candles.

● Chinese New Year celebrations can last up to 15 days. They begin between the middle of January and the middle of February.

Who brings in the New Year with a bang?

New Year in China starts with beating drums, crashing cymbals, fireworks, and a lion or dragon dancing through the streets. All this noise is meant to chase away the bad days of the past and bring luck in the future.

● The Hindus call their New Year festival Diwali. Diwali falls not in January, but in October or November.

● On New Year's Eve in Ecuador in South America, people burn the Old Year on a bonfire! It's a figure they make from bits of straw.

Who blows a horn at New Year?

The Jewish New Year festival of Rosh Hashanah begins with the blowing of a curly ram's horn. The sound calls people to the synagogue to pray for God's forgiveness for the things they've done wrong in the past year. As the new year begins they can make a fresh start.

● Rosh Hashanah falls in October or November. People eat apples and bread dipped in honey in the hope of a sweet year to come.

How long is a month?

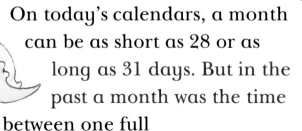

On today's calendars, a month can be as short as 28 or as long as 31 days. But in the past a month was the time between one full moon and the next. Every month was the same length and that was 29½ days.

● The Moon is a dark, lifeless world. It looks bright because one side is always lit up by the Sun. What we call moonlight is really reflected sunlight.

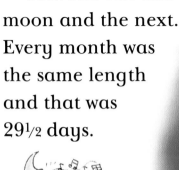

●A piece of music by the German composer Beethoven was named the "Moonlight Sonata." You can imagine the Moon rising as you listen to it.

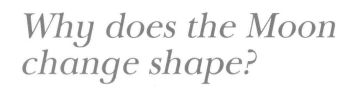

Why does the Moon change shape?

The Moon doesn't really change shape. What changes is the shape of the sunlit part of the Moon that you can see from Earth. As the Moon moves around the Earth in its orbit, the Sun lights it from different directions. First the bright side seems to grow, and then it seems to shrink.

FULL MOON

● The first man on the Moon was Neil Armstrong, a U.S. astronaut who took part in the Apollo 11 space mission in July 1969.

Who eats the Moon?

Chinese children eat delicious moon-shaped cakes on the Moon-viewing festival. At the September full moon, families walk to the park carrying lanterns. Then they eat their cakes and admire the Moon!

Why are there seven days in a week?

● We split the week into work days and rest days. After five hard days at school, everyone loves the weekend!

No one's quite sure how there came to be seven days in a week. Once, it may have been the time between one market day and another, or maybe it was just a quarter of a moon month. But after all this time, a seven-day week feels just about right.

When is it 13 o'clock?

After 12 o'clock, of course! Clocks that number the hours from 1 to 24 are known as 24-hour clocks. Until noon, the hours are numbered 1 to 12. After noon, we just keep on counting.

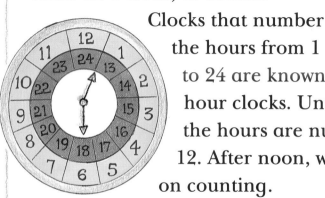

● The ancient Greeks had a ten-day week. It was a long wait until the weekend!

How long is a minute?

A minute isn't long—it's just enough time to peel an apple! There are 60 seconds in a minute, and 60 minutes in an hour. The number 60 was chosen over 5,000 years ago, probably because so many different numbers will go into it evenly.

● The first clocks and sundials showed only the hours. Nowadays people need to know the time to the nearest minute—to catch trains on time, for example!

Why does my watch tick?

More than 20 tiny wheels are packed neatly inside a windup watch. One part inside the watch goes back and forth to make the ticking sound and turn the wheels. The moving wheels keep time and slowly turn the hands around the watch face.

How did people manage before clocks?

Before there were clocks, people judged the time by looking at the Sun. They got up at sunrise and went to bed when it was dark. They ate lunch when the Sun was high overhead, and ate dinner when it set in the west.

How does a grandfather keep good time?

A grandfather clock has a long pendulum that swings back and forth in a steady rhythm. With every swing, wheels inside the clock slowly turn, moving the hands around the face. Winding the clock with a key stops it from slowing down.

● Sundials are one of the oldest kinds of clock. Instead of a moving hand they have a shadow, cast by the Sun. As the Earth turns during the day, the "hand" moves around the clock.

How can you split a second?

Today's electronic timers are so accurate that they can split a second into a million parts. In competitive events, athletes are timed to a hundredth of a second— less than the time it takes you to blink.

Which timer was a load of old rope?

About 400 years ago, a ship's speed was measured with a log tied to a knotted rope. The sailors threw the log overboard, and counted how many knots unwound as the ship moved forward. They used an hourglass to time it accurately. Sailors still measure a ship's speed in knots. One knot is just over one mile an hour.

• Musicians use noisy clicking timers called metronomes to help them keep time when they practice. In an Orchestra they have a silent, waving conductor

How can you cook an egg with sand?

Four minutes is all it takes for the sand inside an egg timer to run from the top to the bottom. And that's just the right time to boil an egg. Sandglasses are simple and accurate and have been used for hundreds of years. Once the timer has finished, you just turn it upside down to start again!

• Speedometers are attached to the wheels of bicycles to measure how fast the bike is going.

What's the time on Earth?

What time it is depends on where you are. At exactly the same moment, clocks around the world will tell completely different times. Every country sets its own time so that it's noon when the Sun is highest in the sky.

Alaska, U.S.A.
It's 7 o'clock. It's the start of a new day.

New York, U.S.A.
It's noon—time to think about lunch!

● Traveling across time zones can confuse your body clock. Air passengers on concorde may eat breakfast in London, fly to New York, and arrive there just in time for...breakfast!

28

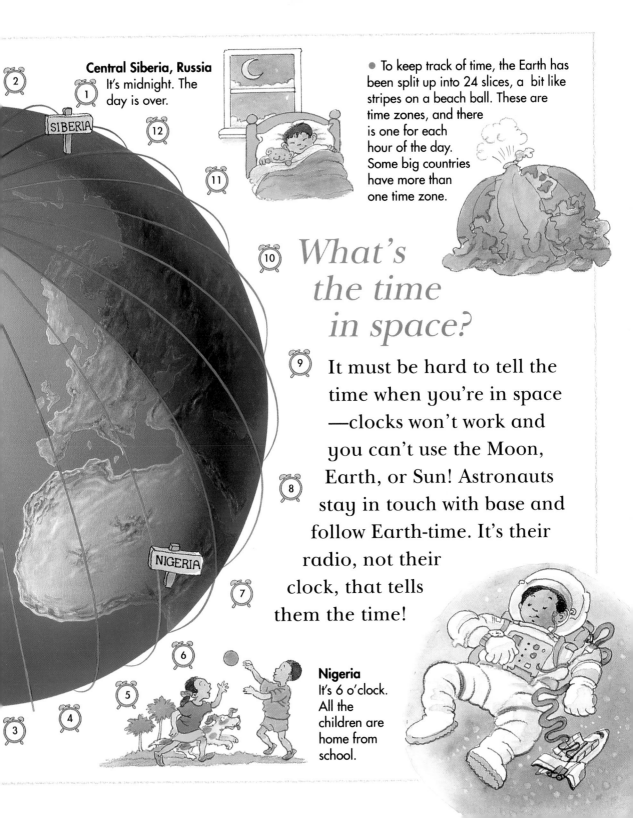

Central Siberia, Russia
It's midnight. The day is over.

● To keep track of time, the Earth has been split up into 24 slices, a bit like stripes on a beach ball. These are time zones, and there is one for each hour of the day. Some big countries have more than one time zone.

What's the time in space?

It must be hard to tell the time when you're in space —clocks won't work and you can't use the Moon, Earth, or Sun! Astronauts stay in touch with base and follow Earth-time. It's their radio, not their clock, that tells them the time!

Nigeria
It's 6 o'clock. All the children are home from school.

How long does a person live?

● Most women live longer than men. The oldest person ever is a woman who has lived for over 120 years.

Most of us will live to enjoy about 70 birthdays—as long as we stay fit and healthy. Humans live about the same length of time as elephants, ravens, and some parrots!

● King Tutankhamen of Egypt died when he was only 18 years old, but his tomb has survived for over 3,000 years.

How long does a mayfly live?

A mayfly usually only lives for a day. It unfolds its wings for the first time in the morning and folds them for the last time at night. That's just long enough for mayflies to mate and lay their eggs before they die.

● Cats live for about 15 years. Mice only live for two—that's if they're not caught first!

How long does a tree live?

Trees grow very slowly and have long lives. Most trees live for between 100 and 250 years, but some bristlecone pines are over 4,500 years old. They are some of the oldest living things on Earth.

● Trees make new wood every year and leave telltale rings in their trunks. Counting the rings on a tree stump tells you how old the tree was when it was cut down.

Index